Thomas Kingsmill Abbott

Elementary Theory of the Tides

Thomas Kingsmill Abbott

Elementary Theory of the Tides

ISBN/EAN: 9783744679459

Printed in Europe, USA, Canada, Australia, Japan

Cover: Foto ©Andreas Hilbeck / pixelio.de

More available books at **www.hansebooks.com**

OF

THE TIDES:

THE FUNDAMENTAL THEOREMS DEMONSTRATED
WITHOUT MATHEMATICS,

AND THE

INFLUENCE ON THE LENGTH OF THE DAY DISCUSSED.

BY

T. K. ABBOTT, B.D.,

FELLOW AND TUTOR, TRINITY COLLEGE, DUBLIN.

LONDON:

LONGMANS, GREEN, & CO., PATERNOSTER-ROW.

1888.

DUBLIN:

PRINTED AT THE UNIVERSITY PRESS,

BY PONSONBY AND WELDRICK.

PREFACE.

———◦◦◦———

THE substance of the following pages has already appeared, partly in the *Philosophical Magazine*, 1871, 1872, and the *Quarterly Journal of Mathematics*, 1872, and partly in *Hermathena*, 1882. Hitherto correct statements about the Tides have been confined to treatises which employ the resources of the higher mathematics. Other works almost without exception* repeat such erroneous statements as that the place of high water without friction would be under the moon, and that high water is retarded by friction. No apology then is needed for the publication in a more accessible form of the present Essay, in which the fundamental theorems are deduced from elementary physical principles without the use of mathematics, except for quantitative calculations. The problem of the influence of the Tides on the length of the day is discussed in a similar method.

---

* The only exception with which I am acquainted is Stubbs' edition of Brinkley's Astronomy, in which the reasoning of this Essay is adopted.

For the benefit of readers who may wish to see the latter problem analytically treated, I have given in an Appendix the substance of Sir George Airy's investigation.

ELEMENTARY THEORY OF THE TIDES.

THE tide-producing force is the difference between the
attraction of the moon (or sun) on the solid body of the
earth, which is the same as if it were all concentrated at
the centre E, and that on the particles of the ocean at x, z.

Confining ourselves to the moon :—

*The direction of the tide-producing force is always tangen-
tial, and towards the line joining the centres of the earth
and moon.*

First, the vertical component being in the same line
as gravity (either in the same or an opposite direction)
cannot directly produce any motion.   In fact, it could not
do so unless it actually exceeded the force of gravity.
And it is too minute to produce any indirect effect.

Secondly, the tangential component is the difference
between the tangential components of the moon's force at
the centre and at the surface.   Now (see figure), at a point
x in the hemisphere nearer the moon, the force is greater
than at E, and, moreover, makes a less angle with the
tangent ; therefore the effective difference is in the direction of its tangential component, *i. e.* towards C.   At a
point z in the further hemisphere the force is less than
at E, and also makes a greater angle with the tangent ;

B

therefore the effective difference is in the direction opposite to its tangential component, *i.e.* it acts towards *A*. The tide-producing force then always acts towards *EM* (in the direction of the arrows). From this we can deduce theorems relating to the place of high and low water, &c., without requiring to determine the magnitude of the force which will be hereafter taken into account. At present we need only observe that it is very small compared with gravity.

First, then, let us consider the case of water limited to an equatorial canal. The moon being supposed in the equator, we shall establish the following theorems :—

I. If there were no friction it would be low water under the moon, and high water in quadratures.

II. Friction accelerates the times of high and low water.

III. In addition to the oscillatory motion of the water there is a constant current produced by the action of the moon.

IV. The effect of friction on this is to increase the length of the day.

I.—*Without friction it would be low water under the moon, and high water in quadratures.*

I suppose the moon to be fixed, and the earth rotating in the direction *ABCD*, carrying the ocean with it.

Now, in the course of one lunar day every particle of the ocean is subjected to precisely the same forces, acting in the same order of succession and for the same periods, being accelerated for about one quarter of a day, viz.

while passing from B to C; then retarded for a quarter, from C to D, and so on. The variation in the amount of the force does not concern us, being the same for every particle.

This being so, it is obvious that those particles will be moving faster which have been for a longer time acted on by an accelerating force, and the velocity will be a maximum when the accelerating force has acted during its full period, viz. through one quadrant. On the other hand, those particles will be moving slower which have been longer acted on by a retarding force, and the absolute velocity will be a minimum when the retarding

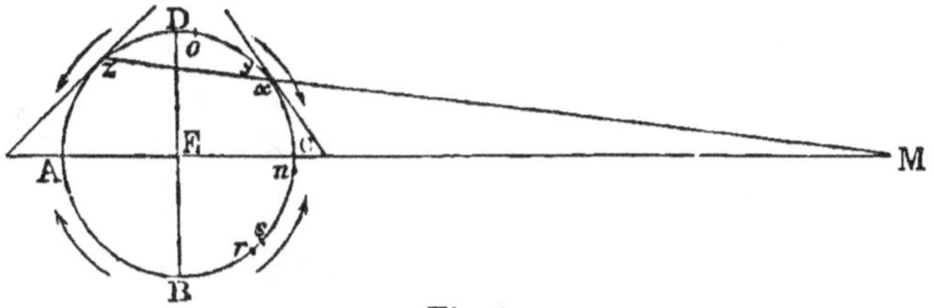

Fig 1.

force has acted during its full period, or through one quadrant. The maximum velocity is therefore at A and C, the minimum at B and D.

Secondly, it is clear that the tide will be rising where each portion of water is moving faster than that just in advance of it; or, in other words, where water is flowing in faster than it flows out. Where this process has gone on for the maximum time the tide will be highest. On the other hand, the tide will be falling where the water is moving slower than that in advance of it—or, in other words, is flowing out faster than it flows in. Where this has continued for the maximum time the tide is lowest.

Now consider any point s in the quadrant BC. The water now passing s has been subject to an accelerating force during the whole time since it passed B, longer therefore than any particles behind it, as at r. It is therefore moving faster; and as the water in the space $r\,s$ is thus flowing out faster than it flows in, the tide is falling. This is the case through the whole quadrant BC.

At C the force changes and becomes a retarding force. The particle at y has been subject to this retarding force longer than one behind it, as at x, and is therefore moving slower. Here, therefore, water is flowing in faster than it flows out, and the tide is rising; and this holds through the quadrant CD. What is said of these quadrants holds also of those opposite to them: the tide is falling all through DA and rising through AB. Hence it is highest at B and D, lowest at A and C. Where will the tide be falling fastest? Clearly where the difference of velocity between r and s is greatest, $i.\,e.$ where the amount of force to which the water at s has been subject since it passed r is greatest—in other words, where the force is at its maximum, viz. at f (fig. 2), 45° from C. Similarly it will be rising fastest at that point in the quadrant CD where the force is greatest, viz. at g, 45° from C.

On the whole, then, the water in the supposed equatorial canal assumes the form of an ellipse; and as it is the earth that is rotating, this ellipse does not change its absolute position except with the moon's motion; only the water accompanying the rotating earth moves fastest at A and C, and is there lowest; and slowest at B and D, and is there highest. Relatively to the earth it is moving westward from e to f and from g to h; eastward from f to

g and from h to e. At A, B, C, D, the particles are in their mean places; at e, f, g, h, they are farthest from their mean places, and change the direction of their relative motion. This is represented in fig. 2, where the inside

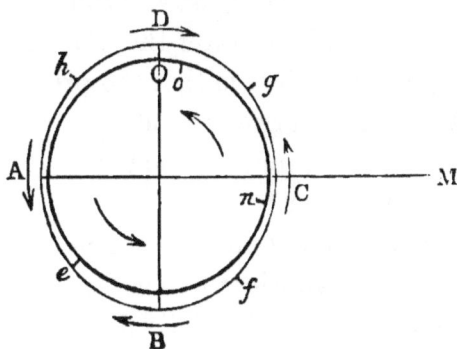

Fig. 2.

arrows show the direction of the earth's motion; the outside arrows that of the relative motion of the water. The path of any one particle may be represented by fig. 3, where the letter A indicates the position of the particle when its mean place is at A in fig. 2.

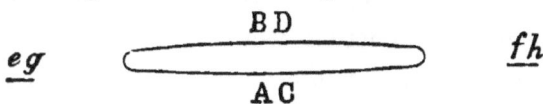

Fig. 3.

II.—*Friction accelerates high and low water.*

The theorem that the effect of friction is to accelerate the time of high and low water admits of an equally simple proof. As the water approaches C, the tangential force diminishes gradually to zero at C. Therefore it must have been equal to the force of friction at some point n (fig. 1), after which friction prevails and the velocity diminishes. It is therefore low water at *n*. Approaching *D*, the ocean is moving slower than the earth; therefore here friction tends to accelerate it, while the retarding force is decreasing to zero. The two forces, then, must be equal at same point *o*, after which the velocity again increases. It is high water therefore at *o*.

It is proper to observe that the preceding proof assumes that the ocean is carried round by the earth in its rotation. This amounts to supposing that it has not assumed a position of equilibrium.*

It is *a priori* an admissible suggestion that the ocean is in a state of equilibrium under the moon's action, *i.e.* that it is absolutely at rest (relatively to the moon), while the earth rotates. But this would imply an apparent movement of the whole body of water with a velocity equal and opposite to that of the earth's rotation, *i.e.* at the equator there would be an apparent current of about 1000 miles per hour. As this does not correspond to the fact, the hypothesis is practically inadmissible; but when friction is considered it appears theoretically inadmissible also. For in this case friction would continually act in the same direction, and its effect would be to make the east-

---

* It is important to observe that we are not entitled to assume that when the tide is rising fastest the water is flowing in from both sides. This is by no means evident. The rate of rise depends on the difference in velocity between two successive parts of the ocean, and this may be greater when the two velocities have the same sign than when they have different signs. Taking into consideration the rotation of the earth, the assumption amounts to this—that the tide is rising fastest where the velocity of the ocean is just equal to that of the earth. This is certainly not evident: in fact it would not be true if the tangential force did not decrease at the same rate on both sides of each of the four maxima. It ought not, therefore, to be assumed, but deduced.

ward forces preponderate; so that although the ocean should be supposed at rest at first, it would ultimately be dragged round by the earth. The actual form of the earth, moreover, in which the equatorial ocean is interrupted by continents, would render this equilibrium of the ocean impossible.

III.—*There is a constant current westward produced by the moon's disturbing force.*

This occurs from two causes. First, the water in the supposed equatorial canal has now taken the form of an ellipse; and, in consequence of friction, the places of greatest elevation are not at B and D, but somewhere in the quadrants BA, CD. Now, the moon's tangential force is, *cæteris paribus*, proportional to the distance of the particles attracted from the centre of the earth. It follows that it is greater in the quadrants BA, CD, than in the other two; but in the former the force is retarding; in the latter it is accelerating; therefore the retarding force exceeds the accelerating, and produces a permanent westward motion.

Secondly, the water having reached its mean place at n, and passed it with its greatest eastward velocity, it is, when it reaches C, eastward of its mean place, *i.e.* it is nearer to g. On the whole way before reaching g it is nearer to that point than if there were no friction; but on passing g it begins to move westward; but its eastward excursion having been shortened by friction, it begins this motion to the west of where it would otherwise be. At o it again arrives at its mean place, which, without friction, it would not reach until D. Thus, in the whole quadrant CD, the particles are nearer to g than if friction had not operated. But the tangential force is greater the nearer the particles are to g, being proportional to sin 2 (angle from moon) = cos 2 (angle from f or g) ; hence the force in the quadrant CD, which is a retarding force, is increased. After passing its mean place at o, the water going westward is, on arriving at D, west of its mean place; and until it reaches h it continues to be west of the place which it would have occupied had friction not operated, $i.e.$ friction withdraws it from h. At h its westward excursion is stopped, and it begins to return eastward. But now from h to A it is eastward of the place due to it without friction. Thus throughout this quadrant the particles are brought farther from h by friction. But here the force is accelerating. Therefore the force in the accelerating quadrants is diminished, while that in the retarding quadrants is increased, and hence again a balance of retarding force, and therefore a current westward. Or thus :—Without friction, the quadrant $f g$, throughout which the water is moving faster than the earth, has its middle point at C; and the following quadrant in which it is slower has its middle point at D. These quadrants are, therefore, equally divided between the accelerating and the retarding quadrants. With friction, the middle points being displaced to n and o respectively, the water is moving faster than the earth through more than half the quadrant BC, and slower through more than half the quadrant CD; and similarly in the opposite quadrants. But BC, DE are the accelerating quadrants, and CD, AB the retarding quadrants. Therefore the water is exposed for a longer *time* to the retarding than to the accelerating force.

We have here, therefore, a *vera causa* which may possibly be effective in retarding the earth's rotation. An attempt will presently be made to estimate the maximum amount of this effect.

On the Quantitative Valuation of the Tidal Disturbance.

Construction for the magnitude of the disturbing force.

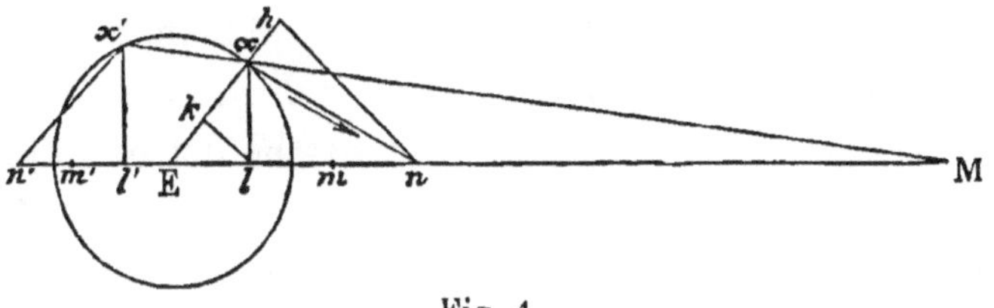

Fig. 4.

Let xl (fig. 4) be perpendicular to EM, and let $lm = mn = El$. Then if ME represent the attractive force of the moon at the centre E, xn will represent the whole disturbing force in magnitude and direction. The proof is as follows:—

Let it be borne in mind that ME is about sixty times the radius of the earth. Hence, if we consider $Ml = Mx$, the error cannot exceed $\dfrac{1}{2 \cdot 60 \cdot 60} = \dfrac{1}{7200}$th part. Again, Mn, Mm, Ml, ME, being arithmetical proportionals with a difference less than $\dfrac{1}{60}$th, may be regarded as geometrical proportionals; the greatest possible error being the same as before. Hence, $Mn : Ml$ or $Mx :: Ml^2$ or $Mx^2 : ME^2$, *i. e.* as the moon's force at E : force at x; therefore, if Mx represent the moon's force at x, Mn will represent the force at E in magnitude and direction, and the difference or disturbing force will be represented in magnitude and direction by xn. In order to have a fixed scale we must represent the force at the centre by ME. On this scale xn is in the nearer hemisphere too small, and in the more remote too large, in the proportion of Mn to ME. This error is at most $\dfrac{3}{60}$ths = $\dfrac{1}{20}$th.

This will be considered by-and-by, but for the present it may be overlooked. The tangential component of xn is equal and parallel to nh, the perpendicular on the radius, and is, therefore, proportional to lk, which is one-third of nh. The same reasoning applies to the dotted letters in the further hemisphere.

To determine more precisely the magnitude of the disturbing force :—

The moon's attraction at x : force of gravity : : $\dfrac{\text{moon's mass}}{Mx^2}$: $\dfrac{\text{earth's mass}}{r^2}$, or nearly $= \dfrac{1}{60^2}$: 87 ; therefore the whole attraction of the moon (represented in fig. 4 by ME) $= \dfrac{g}{87 \times 60^2}$. But on the same scale the greatest tide-producing force is represented by $\dfrac{3}{2} r$ (the greatest value of lk being r), i.e. by $\dfrac{ME}{40}$. (This we shall call H.) The greatest tangential force then is

$$\frac{g}{40 \times 87 \times 60^2} = \frac{g}{12,528,000} = \frac{1}{400,000},$$

nearly. Neglecting the effect of pressure, the effect of the moon's action through one quadrant : the effect of this maximum continued for the same period : : $1 : \frac{1}{2}\pi$ (this appears from the construction in fig. 5).

The number of seconds in a lunar day being 89280, the velocity generated in one-fourth of this time is

$$\frac{22320}{400000 \times \frac{1}{2}\pi} = \frac{1}{28} \text{ feet nearly.}$$

This is the difference between the greatest eastward and the greatest westward velocity; therefore the greatest eastward velocity is $\frac{1}{56}$, and the greatest westward velocity is also $\frac{1}{56}$.

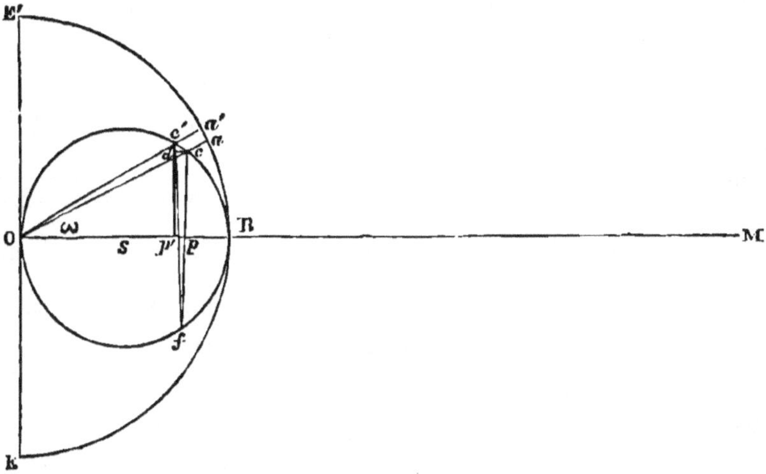

Fig. 5.

As the same amount of water passes through a given section in a given time, the increase in height : total depth of the sea : : relative westward velocity of the water : earth's velocity of rotation (relatively to the moon). The last is about 1486 feet per second. Hence the rise of the tide = $\dfrac{\text{depth of sea}}{1486} \times \dfrac{1}{56} = \dfrac{\text{depth of sea}}{83216}$.

For a sea of three miles in depth this would give 2·268 inches.

The following is a geometrical construction for the velocity and height at any place :—

Round the radius OB describe a circle. Since the angle at BcO is right, it is obvious that Bc is equal to xl in fig. 4, and cp equal to lk; so that the tangential disturbing force at a is proportional to the perpendicular cp. If aa' be the space passed over in the rotation of the earth in one second, the force acting on the water may be supposed unchanged while it passes from a to a'; and its effect during that interval (*i.e.* in this quadrant, the retardation) will also be proportional to cp or its double cf, and to the time: that is, to aa', or the angle at O, aOa'. Calling M the moon's greatest tide-producing force, r the earth's radius, and τ the angular velocity $= \dfrac{2\pi}{\text{seconds in lunar day}}$ $= \dfrac{2\pi}{89280}$, the retardation $= \dfrac{Mcf}{2r} \times \dfrac{aOa'}{\tau}$. Now the angle at $O =$ the angle at f, being in the same circle; and this angle multiplied by $cf =$ the small perpendicular cd, or pp', which is parallel and equal to it. Therefore the whole retardation since leaving B is proportional to the sum of all the abscissæ pp'—that is, to Bp'. It is $\dfrac{M}{2} \dfrac{Bp'}{r} \dfrac{89280}{2\pi}$. This represents the defect from the greatest eastward velocity; and after passing its mean value at the middle point s it represents a velocity which, relatively to the earth, is westerly. The velocity of the current relatively to the earth is represented by ps.

We shall now show that the height of the tide at a' above its lowest point is also proportional to Bp'.

If at any point in the supposed canal a thin section be taken, the quantity of water entering this section in a given time is proportional to the product of the depth and the velocity. If the water flows in a little more rapidly than it flows out, it is clear that the increase in the quantity contained in the section, and therefore the increase in depth, will be proportional to the difference between these two velocities and to the whole depth $\left(\dfrac{\text{diff. of vel.} \times \text{depth}}{\text{length of section.}}\right)$. This holds as long as the change is small compared with the whole depth. If this be supposed uniform throughout the canal, the increase in it (that is, in the height of the tide) at x' is therefore proportional to the retardation; and since the tide began to rise at B, where the velocity began to diminish, it follows that Bp' is also proportional to the height of the tide at a' above its lowest point.

It is easy to deduce from this construction the corresponding formulæ. For, if $OB = r$, we have

$$Bp = r(1 - \cos^2 \omega). \text{ But } ps = \tfrac{1}{2}r - Bp = \tfrac{1}{2}r(2\cos^2\omega - 1) = \tfrac{1}{2}r\cos 2\omega.$$

And since sB is proportional to the mean height, the defect from this height is proportional to ps, and therefore to $\cos 2\omega$.

The effect of pressure with such a tide will be extremely small. As it operates to send the water away from its position of greatest elevation, it will so far assist the moon's force without changing the place of high water.

In the preceding demonstrations we have supposed the water to be limited to an equatorial canal, the moon also being in the equator. It is desirable to consider what modifications will be introduced, first, by supposing the earth to be uniformly covered with water; and, secondly, by taking into account the moon's declination.

It will save repetition if we state once for all certain general principles which we shall have to employ :—

1°. First, suppose an accelerating force acts alternately in opposite directions; the effect (measured by velocity) increases as long as the force acts in either direction; and therefore the velocity in that direction is greatest at the moment that the force changes its direction.

2°. Secondly, the velocity (diminishing under the counteraction of the force) continues to be in the same direction until this counter force has undone all the work accomplished in that direction by the previous force. If the circumstances are alike in both directions, this will be when the force has done half its work. This is precisely the case of the common pendulum.

3°. Thirdly, in the case before us, the water rises when the particles behind are moving faster than those before. The rate of rise is greatest when this difference is greatest ; but as the effect is cumulative, the whole amount of the rise is greatest at the moment when the difference = 0, and is about to change to the opposite.

4°. Fourthly, as in 2°, this difference ceases to increase (*i.e.* is greatest) when the force (or difference of forces) producing it ceases to act ; but it is not reduced to 0 until the opposite force has done half its work. At this moment the accumulation is greatest.

5°. Fifthly, in the case which we are now considering, the effective force depends on the form of the surface, and *vice versâ*. If, then, when this form is spherical the difference mentioned in 3° were always in the same direction, it would continue to act until a certain permanent alteration was produced. If the difference were constant, a state of equilibrium would be attained; but if it alternately increases and diminishes, then the mean form of the surface will be the same as would be produced by a constant force equal to the mean amount of the actual force. The alternate excess and defect of the latter will cause a periodical motion, just as if it were an independent force.*

First, then, the moon being still supposed to be in the equator, let the earth be uniformly covered with water. The tangential force may be resolved into two components—one touching the parallel of latitude (*i.e.* east and west), the other meridional. These may be regarded as giving rise to distinct waves—one east and west, the other north and south.

The actual amount of these forces may be found as follows :—

By the previous construction (fig. 4) (*ME* being moon's force at *E*), the disturbing force at *A* is represented by

$$\frac{3}{2} r \sin 2AM.$$

Resolved along the parallel of latitude, this is

$$3r \sin AM \cos AM \sin \theta.$$

---

* If the reader wishes to apply these considerations to the case of an equatorial canal assumed above, it must be observed that there the elevating force is the excess of easterly force acting on any particles of water above that which affects those in advance, *i.e.* to the east of them. This excess is positive from 45° west of the moon to 45' east (*i.e.* while the moon passes from 45' east zenith distance to 45° west), then negative for 90', and so on.

But by the right-angled spherical triangle (fig. 6)

sin AM sin θ = sin MB (hour angle from moon),

and   cos AM = cos MB cos AB (latitude).

Hence the component parallel to equator

= H cos lat. sin 2 (hour angle)

(H being the greatest disturbing force). This is less than the force in the equatorial canal in the proportion of cos lat. : 1. But the velocity of rotation is less in the same proportion; hence the rise of the tide will be the same. If this force were alone (that is, if the water moved in canals

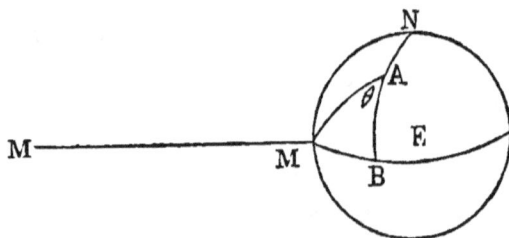

Fig. 6.

parallel to the equator), the ocean in every circle of latitude would take the form of an ellipse with its short axis towards the moon. But these ellipses would not be similar unless the depth of the sea varied as cos lat.

The effect of the meridional component is of a different kind. Its value is

$$\frac{3}{2} r \sin 2AM \cos \theta.$$

But   sin AM cos θ = sin AB cos MB,

and   cos AM = cos MB sin AB (as above).

Therefore this component

$= H$ sin 2 lat. cos^2 (hour angle)

$= \dfrac{H}{2}$ sin 2 lat. $+ \dfrac{H}{2}$ sin 2 lat. cos 2 (hour angle.)

The mean value of this is the first term, the effect of which is to cause a permanent accumulation at the equator.

The second term represents the tide-producing part of the force. This is positive as long as the hour angle from the moon is less than 45° on either side ; and in that case from the equator to lat. 45° this is an elevating force, being greater as the particles are further from the equator : from 45° to the poles it is depressing. In the remaining quadrants this term is negative. Hence, by 5° and 4°, the elevation at the equator (and up to lat. 45°) will be greatest (*i.e.* it will be high water), 90° from the moon. Beyond lat. 45° the depression will be greatest under the same circumstances. In these latitudes, therefore, the effect of the former component would be partially counter-acted. It is easy, however, to see that the variation in the meridional force (and it is only the variation that affects the tide) is in any latitude less than that in the force parallel to the equator in the proportion of sin lat. : 1 ; so that while the height of the tide would be lessened, the place of high water would be as before. The actual magnitude of the tide may be ascertained as follows :—

The force being $\dfrac{H}{2}$ sin 2λ cos $2m$ (λ being lat., and m hour angle).

In order to apply the same method of summing as in fig. 5, we write this

$$\frac{H}{2} \sin 2\lambda \sin 2(45° - m).$$

c

Then, as in fig. 5,

$$\text{velocity} = \frac{H}{4r} \sin 2\lambda \cos 2(45° - m).$$

Now, this increase in the height of the water depends on the difference in velocity at two points, of which the latitude is λ and $\lambda + a$ where a is very small. In fact the difference in the amount of the water entering the section and leaving it is equal to the area of the section multiplied by the difference of velocity, and the decrease or increase of height is equal to this difference of amount divided by the area of the surface, *i.e.*

$$\text{Decrease of height} = \frac{\text{depth} \times \text{increase of velocity}}{ra}$$

(the height increasing when the velocity is diminishing, and *vice versâ*).

But a being small,

$$\sin 2(\lambda + a) - \sin 2\lambda = 2a \cos 2\lambda \,;$$

$$\therefore \text{decrease of height} = \frac{DH}{2r\,ar} \cos 2\lambda \sin 2m,$$

and $$\text{total rise or fall} = \frac{D}{r} \frac{H \cos 2\lambda}{4r^2} \cos 2m.$$

This is $= \cos 2\lambda \times$ half the total rise or fall in the equatorial canal with the moon in the equator.

After passing 45° latitude, the decrease in the circles of latitude requires to be noticed. If we assume our meridional canal to be of uniform width, then the canals will gradually overlap, the tide thus diminishing until at the pole, as is obvious, there will be no tide.

Let us now consider the case of the moon having a declination, which for simplicity I shall suppose less than 22° 30′. This limitation will not affect our results. We shall, as before, take the two components separately.

With respect, then, to the component which acts parallel to the equator. Near the equator itself the considerations previously applied still hold good. Next consider a place a, whose polar distance is less than the moon's declination, to which therefore the moon is circumpolar, and (with the assumed declination) alternately north and

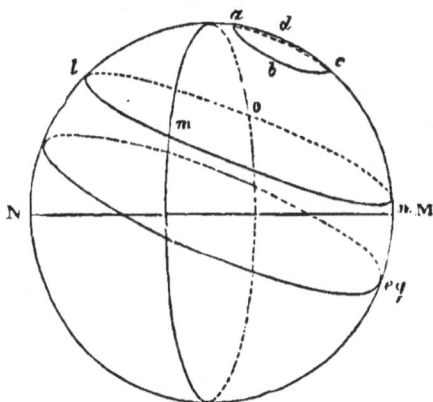

Fig. 7.

south of the zenith. If $abcd$ be the circle of rotation of such a place, it is obvious that the water will be accelerated through the whole semicircle, abc, and retarded through cda. The same reasoning as already employed will show that it will be low water at c and high water at a. Now take an intermediate place whose circle of rotation is $lmno$. Here the water is retarded and rising from l to m and from n to o; and accelerated and falling from m to n and from o to l, and the interval olm is less than mno. Hence the

tide is lowest at *n* and not so low at *l*, and it is high water
at *m* and *o*. Hence we have a diurnal tide in addition
to the semi-diurnal, this diurnal tide becoming of more
importance as we recede from the equator until the co-
latitude = moon's co-declination, when the semi-diurnal
tide disappears.

The meridional component at the equator acts during
half a rotation northward, and during the other half
southward, and in each case is an elevating force, which,
as before, has its greatest effect 90° from the moon. At
all places whose latitude is less than the moon's declination
there is a permanent accumulation. In the circle *abcd*
this component is directed towards the north at *a* and
towards the south at *c*, the points of change being where
the great circles from *M* touch *abcd*. This gives rise to
a north and south oscillation. The southerly force being
the greater, there will be a residual depression of the water
in this region. The depressing force, however, varies,
being greatest at *a* and at *c**, while the elevating force
is greatest where the tangents from *M* meet the circle.
Hence, by 4° and 5° the tide will be lowest at the latter
points and high at the former, and there will be a diurnal
tide, as in the former case. Combining this with the
former result, the effect of both components together will
be to give high water at *a*.

It is not necessary to enter into a detailed examination

---

* If the moon's declination were greater than 22° 30′, *c* might be less than
45° from *M*, in which case the force there would be an elevating one.
Again, at a place whose latitude was greater than 22° 30′, and less than the
moon's declination, the moon's least nadir distance (= *lN*) would be greater
than 45′, and the force depressing.

of the state of things at intermediate places. It is not difficult to see that, as long as the moon's declination is small, there will be an accumulation effected by the meridional component extending from the equator to about lat. 45°, and that, as the moon's declination increases, the accumulation becomes less at the equator and greater towards 45°. If the declination were exactly 45°, there would be no accumulation at the equator, but two elevated rings at lat. 45°. With a greater declination these rings would approach the poles; and obviously, if the moon were at the pole, the ocean would take the form of a prolate spheroid.

The place of high water at any latitude, as far as this is due to the meridional component, would be easily found; but the proportionate effect of the meridional and equatorial components depends partly on the latitude and partly on the moon's declination; and it does not come within the scope of the present essay to solve this problem. It is sufficient to observe that the importance of the meridional component increases with the declination as well as with the latitude. If the moon were at the pole this force would be alone; and, whatever the declination, it alone produces an effect at the pole.

The same reasoning applies, *mutatis mutandis*, to the solar tide.

It was remarked, on p. 10, that the disturbing force is slightly greater on the side nearer the moon than on the remoter side. The effect of this inequality is to produce a small diurnal tide.

ON THE EFFECT OF THE TIDES ON THE LENGTH OF
THE DAY.

§ I.—*Historical.*

In the year 1754 the Berlin Academy proposed, as the subject for a prize essay, the question, " Does any Cause exist tending to Retard the Rotation of the Earth?" What the result of the competition was I do not know; but the question led to the publication by Kant of a short essay, in which he suggested that such a retarding cause existed in the tides. He worked out this suggestion in a rough way, there being, as he truly said, no ascertained data on which any trustworthy calculation could be built.

Laplace examined the question from the historical side, with the help of the records of ancient eclipses, and came to the conclusion that the period of rotation had not altered.

Recently, in consequence of the improvement of the lunar tables, astronomers have seen reason to re-open the question. It has been inferred from the records of ancient eclipses that the day is lengthening at the rate of one second in two hundred thousand years. At first sight this may seem to be an amount too small to leave any trace in history. It must be remembered, however, that in calculating what part of the earth's surface came into the shadow of a given total eclipse, say 2500 years ago, we have to "unwind" 2500 times 365 ($= 91,250,000$) rotations, and a difference amounting to an eightieth of a second between the first and last of these would in the whole period have a very considerable effect.[*] M. Delaunay attributes the retardation to the moment of the moon's disturbing force on the tidal prominences. He

---

[*] About 100 minutes: see Ball, "Elements of Astronomy," p. 377.

started from the assumptions that without friction it would
be high water under the moon and anti-moon, and that
friction retards the time of high water. Both these assumptions were erroneous; but they so far counteracted
one another as to leave the place of high water in the
same quadrants as the true theory, viz. in the quadrants
east of the moon and anti-moon, in which the moon's force
is retarding.

Sir George Airy corrected these errors, and working out
the equations, found two terms which indicate a constant
current westward—one term (the smallest) depending on
the vertical, and the other on the horizontal, displacement
of the water.

In my own Essay on the Theory of the Tides (*Quarterly
Journal of Mathematics*, 1872, and *Philosophical Magazine*),
the effect of friction was indicated, but there was no attempt to estimate it quantitatively. I am not aware that
any attempt has been made to solve this problem; and
indeed it would be absurd to pretend to do so with any
degree of accuracy. What I propose to do is to estimate
the effect so far as to enable us to form a judgment as to
the actual importance of the tides as a cause retarding the
earth's rotation.

It will be convenient first to prove the following propo-
sition respecting the effect of obstacles:—

§ II.—*Obstacles which check the motion of the water towards
a certain point retard the time of high water, and
increase the height.*

If the obstacle is a complete barrier, the tide will rise as
long as the motion of the water is *towards* it, and will fall

as long as the motion is *from* it. Hence, at 45° east of quadratures it will be high water on the east of such an obstacle, and low water on the west of it. The influence of this on the time of high water at other places will extend as far as the pressure is felt.

An obstacle not sufficient to stop motion altogether will produce a similar effect, but of course much smaller, in consequence of the continuity of the surface. If the obstacle be such as to destroy half the velocity of the water, then high water would be 30° after quadratures. In both cases the height would obviously be increased.

It appears from this that the effect of such obstacles is in both respects the reverse of that of friction.

§ III.—*Effect of the moment of the moon's attraction on the tidal prominences in an equatorial canal with the moon in the equator.*

This is the way in which the retardation was supposed by Delaunay to be produced, and Thomson and Tait have adopted the same view.[*]

Now, in order to estimate the greatest effect possible, let us suppose that the greatest elevation is in the middle of the quadrant, *i.e.* 45° before quadratures ; and further that the elevation is not diminished by friction.

Let H = the moon's greatest horizontal force.

   ω = angle from the moon.

   e = greatest elevation.

---

[*] The statement that the earth rotates in a " friction collar," which seems to put the matter in a nutshell, obviously assumes that the passage of the tidal wave is the passage of a mass of water. But this is true only so far as there is a residual westward current, which is certainly not self-evident.

Then the tangential force at any point $= H \sin 2\omega$, and the elevation $= e \sin 2\omega$. Multiplying by the element of the equator, we get the moment $= He \sin^2 2\omega \times r d\omega \times r$. The constant part of this is $= \frac{1}{2} H e r^2 d\omega$.

Summing round the circle, and multiplying by the co-efficient of friction, we have for the whole moment $He\pi r^2 f$.

Taking the density of the earth as 5, the moment of inertia of the equatorial section of the earth is $\frac{5}{2} \pi r^4$. Dividing the former by the latter, we have for the angular acceleration $\frac{2He f}{5 r^2}$.

Now, $\quad H = \dfrac{32}{13 \text{ millions}}$ and $e = \dfrac{\text{depth of sea}}{84,000}$.

If we assume the depth of the sea to be 3 miles, the angular acceleration becomes nearly $= \dfrac{f}{203 \text{ billions} \times r}$.

Multiplying by the number of seconds in 100,000 years (about 3 billions), we obtain $\dfrac{1}{68} \dfrac{f}{r}$ nearly.

Now, the velocity of the earth's surface at the equator is about 1530 ; the angular velocity therefore is $\dfrac{1530}{r}$.

Hence, $\qquad \dfrac{1}{68} f = \dfrac{\text{earth's velocity}}{104,040} f$.

If the earth's velocity is diminished in this proportion, the length of the day will be increased by

$$\frac{86,400}{76,500} f \text{ seconds} = \text{nearly } \cdot 83 f \text{ seconds}.$$

Now, in the case supposed, f is excessively small, the friction being chiefly that of water on water. Hence we conclude that in an unobstructed equatorial canal the effect of friction in retarding the rotation would be quite insignificant, even on the supposition above adopted, that the place of high water is 45° before quadratures. If this place were affected only by friction, the displacement would really be only a few degrees, if so much. It appears, therefore, that the direct effect of the moon's disturbing force on the tidal prominences is wholly insensible. It would not amount to one second in a thousand million years.

But there is another way of viewing the matter, which does not introduce f. The following consideration explains this :—

§ IV.—*Of the effect of the residual current westward due to the change in the time of high water.*

The constant force found above = $\frac{1}{2}He$ produces an accumulating westward tendency in the water. This once impressed will continue to increase until checked by friction, that is, until friction becomes equal to this constant force. Therefore when we take a sufficiently long time we may assume that the total moment (of the water) is not affected by the coefficient f—that is to say, on the assumption made above, that the elevation is not affected; and friction being so slight, this may be assumed. Moreover, although friction alone could not accelerate high water so much as three hours (= 45°), our conclusion will hold if the displacement takes place from any other cause.

This being premised, I shall now examine the question from the point of view suggested by Airy.

§ V.—*Effect of the changes in the disturbing force due
to the displacement of the water.*

By substituting, in the expression for the disturbing
force, the altered value of the ordinate of the water for the
original value ($x + X$, for x), Airy finds that the expression contains a constant term dependent on the distance
of high water from quadrature. The source of this constant term may be understood from the following observation :—

The particles are in their mean place at the moment of
high water and at that of low water; at the former they
are travelling W. with their greatest velocity; at the
latter they are travelling E., also with their greatest
velocity. Now, the place of high water being W. of
quadrature, and the water moving W., it follows that
when the water reaches quadrature, approaching the moon,
it is behind, or W. of the place which, without friction, it
would have occupied. On the other hand, at syzygy it is
in advance, or E. of its place. In both cases the disturbing force is diminished by this displacement, the force being
greater the nearer the particles are to the middle point of
the quadrant. In other words, $\Pi \sin 2\omega$ is diminished
throughout, ω being increased when over 45°, and diminished when less than 45°. In the following quadrant,
i. e. after passing the moon, the opposite change takes
place, since the particles enter it E. of the place they
would otherwise occupy, and leave it W. of their place.
Now, the former quadrant is that in which the moon's
force is accelerating, the latter that in which it is retarding.
The same observation applies to the other two quadrants.

Thus the accelerating and retarding forces are no longer in equipoise, the latter predominating.

To calculate the effect:—The maximum excursion of the water without friction in the case of a canal three miles deep would be about 126 feet. Assume that this is undiminished; and assume, as before, that it is high water 45° W. of quadratures. Then we may assume the displacement at each point to be 126 cos 2ω; and the moon's force being H sin 2ω, the change in the disturbing force due to this displacement

$$= 2H \cos 2\omega \times \frac{126}{r} \cos 2\omega \text{ per second.}$$

The constant part of this $= H \dfrac{126}{r}$.

Putting for H its value $\dfrac{1}{400,000}$, and calculating the effect continued for one lunar day (89,000 seconds), we have

$$\frac{10}{45} \times \frac{126}{r}, \text{ or } \frac{100}{75 \text{ millions}}, \text{ nearly.}$$

This acts on the whole mass of the canal. Introducing the moments, as in p. 25, we have as the acceleration for one day

$$\frac{200}{75 \text{ millions}} \times \frac{\text{mass of canal}}{\text{mass of equatorial section of earth} \times r}.$$

With the assumed depth of sea, the latter factor $= \dfrac{1}{3300 \times r}$.

Hence the daily angular acceleration

$$= \frac{1}{375,000} \times \frac{1}{3300 \times r}.$$

Multiplying by the lunar days in 100,000 years (about 33 millions), we have as before $\dfrac{1}{68r}$. This gives a retardation of about ·83 seconds.

For the reason before stated, it is unnecessary to multiply this by the coefficient of friction.

There is a third way of viewing this cause. Owing to the displacement of the place of high water, since that is the point where the water is moving fastest westward, the water is a longer time in the retarding quadrants than in the others, e.g. on the previous hypotheses it is 126 feet behind its place on entering the accelerating quadrant, and 126 feet in advance on leaving it. It is, therefore, in that quadrant about $\dfrac{252}{1480}$ seconds = about ·17″ less than a quarter of a lunar day. This would give a similar result to that already found.

We have yet to take into account the solar tide. Reckoning this as $\dfrac{22}{50}$ of the lunar, we have as the maximum total effect 1·2 seconds. If we suppose the actual effect to be half the possible maximum, this would agree curiously with the result deduced from eclipses, viz. 1·0 second in 200,000 years. But too much importance must not be attributed to this coincidence.

The preceding calculations are obviously applicable to the case of a globe uniformly covered with water, since each section parallel to the equator would give the same results. The meridional wave would have no effect on the rotation.

It is not worth while to extend our calculation to the case of the moon not being in the equator. The nett result would be to diminish the retardation.

§ VI.—*Application to the actual state of the earth's surface.*

In attempting to apply the preceding results to the actual condition of things on the earth's surface, the following points must be noted:—

First.—On the earth as it actually is the effect of friction proper on the tides is trifling compared with that of obstacles. Against these the tidal current impinges, and in addition the increased elevation gives the moon an increased pull, which, if acting towards the obstacle, exerts its full moment on the earth, but only for a fraction of a day.

Secondly.—The existence of a retarding influence depends, as we have seen, on the place of high water being in what I have called the retarding quadrants, *i.e.* less than six hours in time later than the moon's meridian passage. If this condition is violated, the influence might be accelerating. Suppose a continent whose coasts run N. and S. (as those of America may be roughly said to do); then if it is high water on the east coast less than six hours after the moon, the effect of the pull just mentioned is retarding; if it is high water on the west coast more than six hours after the moon, the effect is to accelerate. In other cases the direct effect is *nil*.

Now, owing to the great irregularity of distribution of land and water, theory will not help us in determining the times of high water; but on consulting the Tables founded on observation we find, for example, the following results:—

In the open part of the Pacific Ocean high water is about 30° before quadratures; farther from the equator at both sides it is at quadratures; farther still it is 30° after quadratures.

(Confining ourselves to the direct effect of the moon's action on the tidal prominences)

The effect on—

East coast of China, ...	None.
,, India, ...	,,
,, Australia, ...	,,
,, Africa, ...	Retarding.
,, S. America, ...	,,
,, N. America, ...	None.
West coast of S. America (Peru), . .	Accelerating.
,, N. America (California, &c.), .	,,
,, Australia, ...	,,

(It should be observed, that for the present purpose we ought to take the time of high water where the depth of the sea begins sensibly to diminish in approaching the coast, but this we are unable to do.)

These instances are sufficient to show the difficulty (perhaps amounting to impossibility) of determining whether there is any preponderance of retardation at all. At all events, however, it is clear that the retardation, if any, must fall very far short of the maximum.

It is to be remembered, further, that in the case considered above of a globe uniformly covered with water, each section of the globe parallel to the equator has its own tidal current to encounter its own inertia, and hence the result in the case of the equatorial canal was applicable to the entire globe. But in the case of the earth it is not so, and this would still further diminish the retardation.

On the whole, it would appear that no certain conclusion as to the amount of retardation of the earth's rotation by the tides can be drawn from theoretic considerations.

APPENDIX.

SIR GEORGE AIRY ON THE TIDES.*

THE case considered is that of the water in an equatorial canal of uniform depth. Adopting some one point of the canal as zero of measure, let x, measured westwardly from that point, be the abscissa for any point of the fluid under consideration; y the similar abscissa for the point to which the moon is vertical. If r be the earth's radius, the angular distance of the moon westward from the meridian of the point x will be $\frac{y}{r} - \frac{x}{r}$. Then it is known from the ordinary theory of perturbation that the horizontal force produced by the moon on particles in the place x and its neighbourhood may be represented by $\Pi \sin 2\left(\frac{y}{r} - \frac{x}{r}\right)$. The measure y is proportional to the time, and therefore the expression of this force may be put in the form $+ \Pi \sin (it - mx)$, the positive sign being used; because when $\sin 2\left(\frac{y}{r} - \frac{x}{r}\right)$ is positive, the force tends to move the water in the direction in which x is measured.

A second force is derived from friction. Suppose that the mean abscissa for any particle is x, but that its true disturbed

---

* *Monthly Notices* of Royal Astronomical Society, 1866, p. 221.

abscissa at any moment is $x + X$ (X depending on, or being a function of, x and t). Then the velocity of the particle is $\dfrac{dX}{dt}$. And, supposing the friction proportional to the velocity, and the direction always opposed to the direction of motion, it may be represented by $-f\dfrac{dX}{dt}$.

A third force depends on the form of the surface of the water. Let the mean depth of the water be k. Then, in the mean state of the water, the volume included between the points whose mean ordinates (abscissæ) are x and $x + \delta x$, is $k\delta x$. But in the disturbed state x is changed to $x + X$, and $x + \delta x$ is changed to $x + \delta x + X + \dfrac{dX}{dx}\delta x$. The distance between the points is now $\delta x\left(1 + \dfrac{dX}{dx}\right)$; and as the volume $k\delta x$ is still included between them, the depth of the water now is

$$\frac{k}{1 + \dfrac{dX}{dx}} = k\left(1 - \frac{dX}{dx}\right)$$

nearly, or its surface is raised by $-k\dfrac{dX}{dx}$ nearly. This is the tidal elevation of the point whose mean abscissa is x. The tidal elevation of the point whose mean abscissa is $x + h$ will be

$$-k\left(\frac{dX}{dx} + h\frac{d}{dx} \cdot \frac{dX}{dx}\right), \quad \text{or} \quad -k\frac{dX}{dx} - kh\frac{d^2X}{dx^2}.$$

The excess of the former is $+ kh\dfrac{d^2X}{dx^2}$.

This is the height of a head of water which acts horizontally upon the whole depth k of the water, and of which, therefore, the entire pressure is $+ k^2h\dfrac{d^2X}{dx^2}$. The volume of water on which it acts is kh. Hence, by the usual rule connecting pressure with accelerating force, the accelerating force depending on this cause is $+ gk \dfrac{d^2 X}{dx^2}$.

Collecting these three accelerating forces, and forming the usual equation of motion, and remarking that the abscissa to which the motion really applies is $x + X$, but that, as x is independent of t, the expression

$$\frac{d^2(x + X)}{dt^2} \quad \text{is really} \quad \frac{d^2 X}{dx^2} ;$$

$$\frac{d^2 X}{dt^2} = + H \sin (it - mx) - f \cdot \frac{dX}{dt} + gk \cdot \frac{d^2 X}{dx^2}.$$

It is impossible (in our present state of mathematical knowledge) to give a general solution of this equation, and it would be useless if we could give it, because it must include every ripple on the sea. But a solution of the utmost generality in its reference to the periodic forces which produce the tides will be obtained by the following assumption of corresponding periodic character :—

Let $\qquad X = A \cdot \sin (it - mx) + B (\cos it - mx),$

and substitute this for X in the left-hand term, and in two of the right-hand terms of the equation we obtain immediately,

$$0 = (i^2 A + H + fi B - gkm^2 A) \sin (it - mx)$$

$$+ (i^2 B \qquad - fi A - gkm^2 B) \cos (it - mx).$$

And as each of those terms must separately $= 0$,

$$0 = (i^2 - gkm^2) A + fi \cdot B + H,$$

$$0 = (i^2 - gkm^2) B - fi \cdot A ;$$

from which

$$A = - \frac{i^2 - gkm^2}{(i^2 - gkm^2)^2 + f^2 i^2} H, \quad B = - \frac{fi}{(i^2 - gkm^2)^2 + f^2 i^2} H ;$$

and

$$X = - \frac{H}{(i^2 - gkm^2)^2 + f^2 i^2} \{(i^2 - gkm^2) \sin (it - mx) + fi (\cos it - mx)\}.$$

If the constant angle F be determined by the equation

$$\tan F = \frac{fi}{i^2 - gkm^2};$$

where, with a positive denominator, F will always be positive and less than 90°; then

$$\cos F = \frac{i^2 - gkm^2}{\sqrt{\{i^2 - gkm^2)^2 + f^2 i^2\}}}, \quad \sin F = \frac{fi}{\sqrt{\{(i^2 - gkm^2)^2 + f^2 i^2\}}},$$

and the expression for X becomes

$$X = \frac{-H}{\sqrt{\{i^2 - gkm^2)^2 + f^2 i^2\}}} \cdot \{\sin (it - mx) \cdot \cos F + \cos (it - mx) \sin F\}$$

$$= \frac{-H}{\sqrt{\{(i^2 - gkm^2)^2 + f^2 i^2\}}} \cdot \sin (it + F - mx).$$

The tidal elevation of the surface of the water, or $- k \dfrac{dX}{dx}$ is

$$= \frac{-mHk}{\sqrt{\{(i^2 - gkm^2)^2 + f^2 i^2\}}} \cos (it + F - mx).$$

If there were no friction, $f = 0$, $F = 0$, and we should have

$$X \text{ without friction} = \frac{-H}{i^2 - gkm^2} \sin (it - mx).$$

Tidal elevation of surface without friction

$$= \frac{-mHk}{i^2 - gkm^2} \cos (it - mx).$$

In order to arrive at a proper understanding of the import of these expressions, it is necessary first to ascertain whether gkm^2 is greater or less than i^2.

Now, it has been assumed that y is proportional to the time, say $= nt$, in which assumption it is obvious that n is the number of units of linear measure on the earth's equator over which the moon passes in a unit of time, and it or $\dfrac{2y}{r}$ is therefore $\dfrac{2nt}{r}$, and i is $\dfrac{2n}{r}$, where n exceeds 1000 miles per hour, or 1400 feet per second.

The value of m is $\dfrac{2}{r}$; and gk is the product of 32 by the depth of the sea in feet (always using the foot and the second as the units of measure and time). Thus :

$$i^2 - gkm^2 = \frac{1}{r^2}\,(2800^2 - 128 \times \text{depth of sea in feet}).$$

This quantity will always be positive except the depth of the sea exceed 12 miles ; far exceeding any supposed real depth of the sea. The denominator $i^2 - gkm^2$, therefore, is always to be considered positive.

Now when the moon is vertical over any point x of the canal,

$$y - x = 0, \text{ or } 2\left(\frac{y}{r} - \frac{x}{r}\right) = 0, \text{ or } it - mx = 0.$$

Therefore, if there be no friction, the tidal elevation of the surface $= -\dfrac{m\Pi k}{i^2 - gkm^2}$. This is its maximum negative value. Consequently *it is low water under the moon.*

Also, if the term gkm^2 in the denominator be neglected, *the tidal elevation (which has k for a factor) is proportional to the depth of the sea.*

Theorems equivalent to these were proved by Laplace.

If there be friction, the tidal elevation of the water is $- D \cos (it + F - mx)$, (D being the coefficient above). This quantity has its greatest negative value, or it is low water when $it + F - mx = 0$, or when $t = \dfrac{mx - F}{i}$. If there had been no friction, low water would have occurred when $t = \dfrac{mx}{i}$. The former value of t is the smaller; and therefore *the phase of low water (and consequently the other phases of the tide) are accelerated by the friction.*

The magnitude of the tide is evidently diminished (but constantly, not varying from day to day) by the friction; its denominator being $\sqrt{\{(i^2 - gkm^2)^2 + f^2 i^2\}}$ instead of $i^2 - gkm^2$.

The value of friction upon the water at any place x is $-f \dfrac{dX}{dt}$; and therefore the reaction of the water's friction upon the solid channel is $+f \dfrac{dX}{dt}$, or

$$\frac{-\mathit{\Pi}if}{\sqrt{\{(i^2 - gkm^2)^2 + f^2 i^2\}}} \cos(it + F - mx).$$

To obtain the effect of this on the entire solid globe, we must integrate

$$\cos(it + F - mx), \text{ or } \cos\left(it + F - \frac{2}{r}x\right),$$

with respect to x, from $x = 0$ to $x = 2r\pi$. This definite integral is obviously $= 0$, and therefore *the friction does not tend at any instant either to accelerate or to retard the rotation of the solid globe.*

Subsequently, however, Sir G. Airy wrote as follows:—

I have at length discovered two terms which appear to exercise a real effect on the rotation of the earth:—

First.—The coefficient of horizontal forces, which I have designated by $\mathit{\Pi}$, is not constant, but is proportional to the distance from the earth's centre. If we take that value which corresponds to half the depth of the water, the true value, or $\mathit{\Pi}'$, may be represented nearly by

$$\mathit{\Pi} \times \left\{1 + \frac{1}{2r} . \text{ tidal rise}\right\};$$

or $\quad \Pi \times \left\{ 1 - \dfrac{1}{2r} \cdot \dfrac{m\Pi k}{\sqrt{\{(i^2 - gkm^2)^2 + f^2 i^2\}}} \right\} \cos(it - mx + F)$,

and the true horizontal force, or $+ \Pi'\sin(it - mx)$, will be $\Pi \sin(it - mx)$

$$- \frac{m\Pi^2 k}{4r\sqrt{\{(i^2 - gkm^2)^2 + f^2 i^2\}}} \times \{\sin(2it - mx + F) - \sin F\}.$$

Giving to $\sin F$ its value, this contains the constant term

$$+ \frac{m\Pi^2 kfi}{4r\{(i^2 - gkm^2)^2 + f^2 i^2\}}.$$

Second.—The ordinate of the water on which the force is acting is not x but $x + X$; and therefore, instead of using the expression $\Pi \sin(it - mx)$, we ought to use

$\Pi \sin(it - mx - mX)$, or $\Pi \sin(it - mx) - m\Pi X \cos(it - mx)$.

Putting for X its value, the last term becomes

$$+ \frac{m\Pi^2}{\sqrt{\{(i^2 - gkm^2)^2 + f^2 i^2\}}} \cdot \sin(it - mx + F)\cos(it - mx),$$

which, with similar expansion, and substitution for F, gives the constant term

$$+ \frac{m\Pi^2 fi}{2\{(i^2 - gkm^2)^2 + f^2 i^2\}}.$$

This term is much larger than that found above. Call the sum of these two terms $+ c$. Then the equation of motion, as applying to this term only, and putting b^2 for gk, becomes

$$\frac{d^2X}{dt^2} - b^2 \frac{d^2X}{dx^2} = + c.$$

Let $bt + X = u$, $bt - x = v$; then this equation may be changed into

$$4b^2 \frac{d^2X}{du\,.dv} = +c; \quad \frac{dX}{dv} = + \frac{cu}{4b^2} + \phi'(v); \quad X = + \frac{cuv}{4b^2} + \phi(v) + \psi(u),$$

or, $$X = \frac{c}{4b^2}\left(b^2 t^2 - x^2\right) + \phi\left(bt - x\right) + \psi\left(bt + x\right).$$

For determining the form of the arbitrary functions, we remark that the final solution must contain no power of x (otherwise we should have inconsistent values in completing the round of the circle), and that the form of the first term then admits only algebraical functions. Thus we find that the solution must be

$$X = +\frac{c}{4b^2}\left(b^2 t^2 - x^2\right) + \frac{c}{8b^2}\left(bt - x\right)^2 + \frac{c}{8b^2}\left(bt + x\right)^2 = +\frac{c}{2}t^2,$$

$$\frac{dX}{dt} = + ct.$$

That is, there is a constant acceleration of the waters as following the moon's apparent diurnal course. As this is opposite to the direction of the earth's rotation, it follows that, from the action of the moon, there is a constant retarding force on the rotation of the water, and therefore (by virtue of the friction between them) a constant retarding force on the rotation of the earth's nucleus.

If, as in the preceding cases, we include in the equation the term depending on f, the form of solution is somewhat modified, and a better view of the term is obtained. Remarking that, as is stated above, no powers of x can be admitted, we may omit $\frac{d^2 X}{dx^2}$; and the equation as regards the new term becomes

$$\frac{d^2 X}{dt^2} + f\frac{dX}{dt} = + c.$$

The solution of the equation is

$$X = \frac{ct}{f} + c' + c''.\,\epsilon^{-ft},$$

from which the friction

$$= -f\frac{dX}{dt} = -c + f^2 c'' \epsilon^{-ft}.$$

Whatever, therefore, be the primary state of things, the second term will ultimately become insensible; the frictional force on the water will be $- c$ in the direction of the moon's apparent diurnal motion, or $+ c$ in the direction of the earth's rotation; and this implies a force $- c$ in that direction upon the nucleus of the earth, constantly retarding its rotation.

THE END.

Printed by PONSONBY AND WELDRICK, *Dublin.*